CW00494766

Food Activities

Disassembly and Quality Contro

Contents

Pages

The Chase High School,
Geraldine Road,
MALVERN,
Worcs. WR14 3NZ.
Tel: 01684 572096
Fax: 01684 566643

Acknowledgements

My thanks to the following food companies who helped me develop material and information for this book:
Brenda Jamieson of Sainsbury's; Nigel Garbutt, Safeway; Clive de W Blackburn, Leatherhead Food RA, Lorraine Collins, Asda.
To the many teachers who through inservice training days in London, Leeds and Hampshire helped to develop and refine ideas and shared their own teaching experiences.
Thanks to the Committee of London NATHE, especially Maura Gibbons who tested the hot bread snack; Helen Simonds for the Food Specification Worksheet; pupils and staff of Ricards Lodge School, Merton for testing the disassembly of children's food, especially Sue Johne, who I admire for knowing the meaning of the word "didactic". A special thanks to Christine Thompson, Technology Adviser for Hampshire for her up-dates on curriculum and assessment and her enthusiasm for food technology and to Marion Rutland of Roehampton Institute.
To David Smith, technology teacher at Elliott School Putney, for his support and his amusing artwork and page design.
And much appreciation to Mark and to Annabel who has taken a part-time job in a fast food restaurant and taught me lots about quality control, and to Simon who may yet learn to cook for himself.

Food Activities ISBN 09521645 5 8 First published 1995, second publication 1995.

Copyright Ridgwell Press, PO Box 3425, London SW19 4AX. This page may be photocopied by purchasing institution only.

Disassembling food products Written activity

Introduction
This book is witten to help with the National Curriculum Orders for Design and Technology. To develop your **design and technology capability** you will be given opportunities **to design and make products. Food** is one of the design materials. In lessons you will take part in **focused practical activities** in which you develop and practise particular **skills and knowledge**. Activities will help you **investigate, disassemble and evaluate familiar products** such as food products.
The book focuses on the **skills of disassembly and understanding quality.** Disassembly means "taking apart". Food companies often take apart a range of food products when they want to develop new ideas. This book shows you how you can design and make food products by getting ideas from other foods. Start with simple ideas such as muesli, then progress to making more complicated foods as your skills improve.

Milkshake

Meal on a tray

Cake

Sandwich

4 things to find out about each food product

To do
Imagine that you are in a food design team. You have been asked to disassemble the four food products shown above. Decide what information you need to find out for each product. You can look at the label and taste and weigh the product. For each product list four things you can find out about each food which would be useful for designing new products.
Further work Add two food products of your own and explain what you could find out about them to help with design ideas.

Copyright Ridgwell Press, PO Box 3425, London SW19 4AX. This page may be photocopied by purchasing institution only.

Quality of food

Written activity

When we buy any product, we hope it will be of a good quality. Everyone has different views on the meaning of "quality", depending upon our likes and dislikes, and our standards.

To do

Work in groups. Write down the words you would use to describe the quality you would expect if you bought or made one of the following foods:

- a vegetarian lasagne
- a bag of chips
- a salad
- a chocolate milk shake.

Show your words to the rest of the class. Make up your own **Quality Food Word Bank** for use with design work.

Quality Food Word Bank

Looks, taste and nutrition
looks attractive,
good quality ingredients,
right texture,
tasty,
served at right temperature,
healthy,
meets nutritional needs,
fresh,
attractive packaging.

Size, shape and cost
right shape,
right size,
within budget limits,
little waste,
portion size OK,
good value for money,
cheap to buy.

Safety
made in a safe way,
hygienically made,
stored safely,
cooked long enough,
served at the right temperature,
safe to eat,
easy to cook.

Questions

Use the Quality Word Bank to highlight, circle or list words to describe the quality you would like for a) a school meal, b) a sandwich, c) food of your choice.

Copyright Ridgwell Press, PO Box 3425, London SW19 4AX. This page may be photocopied by purchasing institution only.

Getting design ideas from food

Written activity

Disassembly

Disassembly means "taking apart". When you disassemble a food product there are many questions you can ask to get information and ideas for design.

Disassembly of a cake

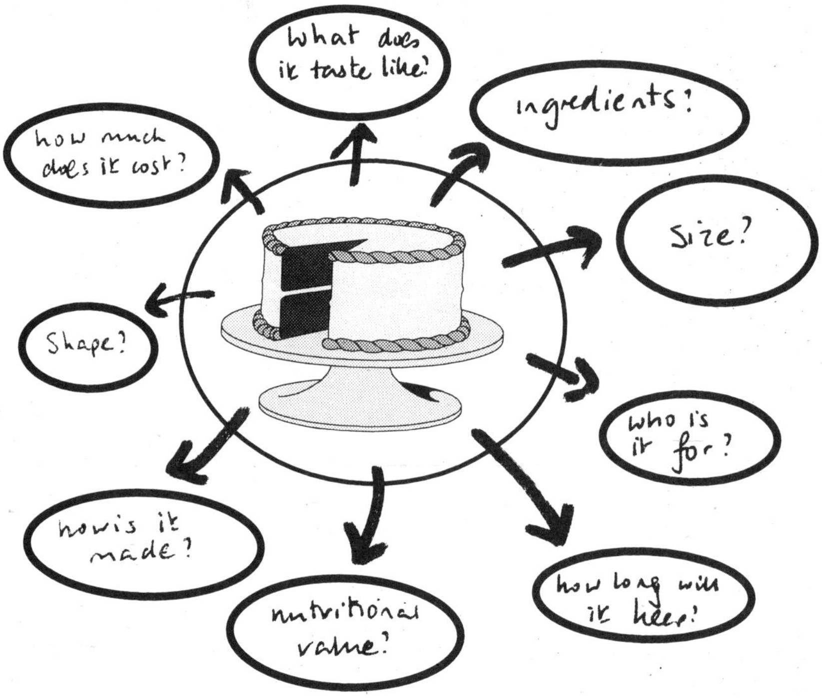

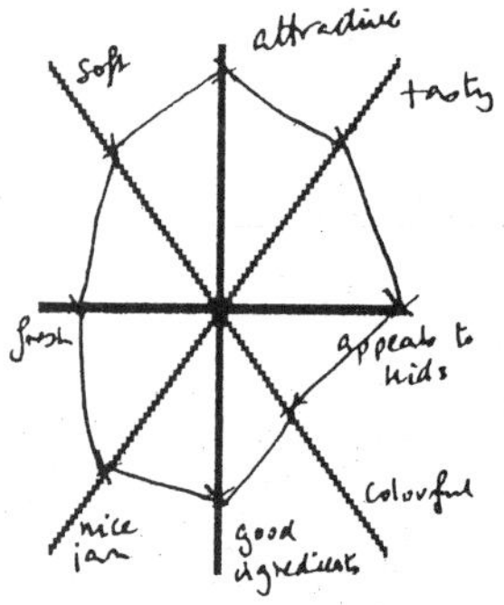

A star profile can show the food product

This information can lead to a specification

> **Specification**
> **Who is it for?**
> A birthday cake for a 6-8 year old
> **What is it made from?**
> Sponge, jam and coloured butter icing
> **Serves** - 10-12
> **Cost to make** - £2.50

Quality needed

Must be safe to eat
Cake must be well cooked and soft
Tasty and colourful icing
Attractively decorated
Must be properly stored
Must look attractive.

> Questions
> 1. Make a list of words to describe how you would like your birthday cake to look and taste.
> 2. How will the cake be made up?
> 3. Draw the type of decoration you would like.

Copyright Ridgwell Press, PO Box 3425, London SW19 4AX. This page may be photocopied by purchasing institution only.

Getting design ideas from food

Written activity

Disassembly

You can get ideas for food design just by looking at food products. Look carefully at these food pictures. Try and work out how they are made, and what they are made from. Complete the chart with your results.

pie

burger

drink

lolly

Food Product Analysis

Fill in the chart for each of the foods shown above. (You could enlarge the chart).

What is it?	pie	drink	burger	lolly
What is it made from?				
How is it made?				
Who would buy it?				

Copyright Ridgwell Press, PO Box 3425, London SW19 4AX. This page may be photocopied by purchasing institution only.

Getting design ideas from food   Written activity

When food companies want to develop a new product, they often look at existing food products. For example, if they want to develop a new range of pasta dishes, the design team goes out and buys a selection of ready-made pasta products. Each product is analysed to find out how it is made and what it tastes like. This process of taking things apart is called **"disassembling products".** The ideas are used for new product design.

To do

You want to design and make a pasta dish. Start by making a collection of ready-made pasta products. Keep the packaging for information. Investigate each product by taking it apart (**disassembly**). You can use any of the charts in this book to help get information.

Make sure you cook and taste each product - food is designed to be eaten!

When the investigation is complete, brainstorm your design ideas for your own pasta dish. Now make it.

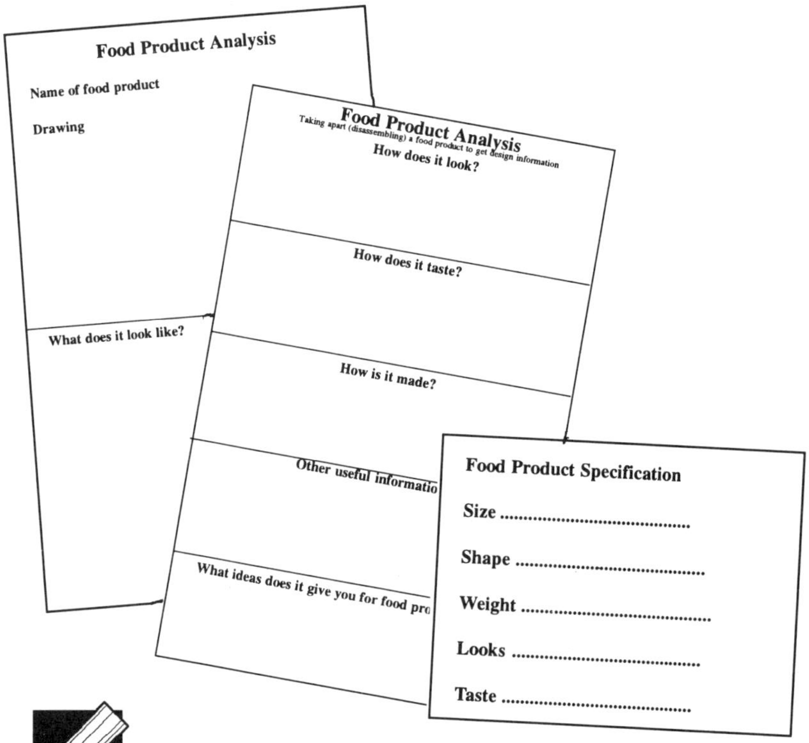

Questions from a packet

The packaging of a food product gives you lots of information.
Use the packet on page 7 to answer these questions:

1. Write down the name of the food product. When can it be eaten?
2. What does it cost, how many does it serve and what does it weigh?
3. How is the product made?
4. How should it be stored?
5. What sort of people would buy this product?
6. How should this product be cooked?
7. Is there a "best before" or "use by" date? Why is this important?

Further questions

1F. What can you find out by looking at the list of INGREDIENTS?
2F. What does the NUTRITION INFORMATION tell you about the product?
3F Where would you write to if there were any complaints?

Design ideas

Make a list of pasta dishes you would like to try and make.

Copyright Ridgwell Press, PO Box 3425, London SW19 4AX. This page may be photocopied by purchasing institution only.

Getting design ideas from food

Written activity

Using a label to get design information

Sainsbury's Vegetable Lasagne is part of Sainbury's Italian Chilled Recipe dish range. Serve as a snack or as a main meal with a mixed salad and hot crusty garlic bread.

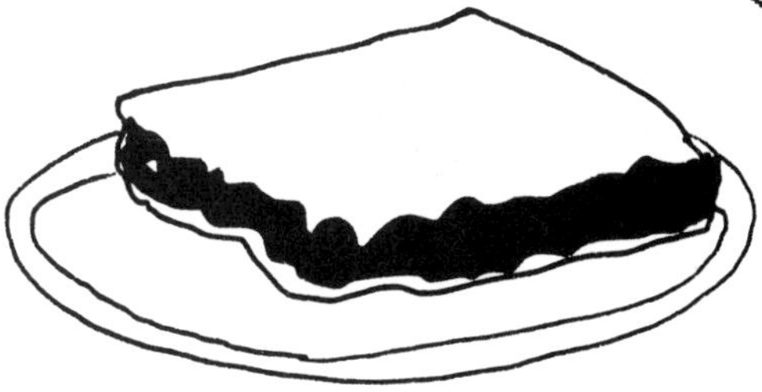

COOKING INSTRUCTIONS

All cooking instructions vary in performance. These are guidelines only. Remove sleeve and lid. Place on a baking tray in a preheated oven 190°C, 375°F, Gas Mark 5 for 35 minutes.

COOKING INSTRUCTIONS FROM FROZEN

Remove sleeve and lid. Place on a baking tray in a preheated oven 190°C, 375°F, Gas Mark 5 for 50 minutes.

NUTRITION	INFORMATION				
TYPICAL VALUES	PER 100g (3.5oz)	PER HALF PACK	TYPICAL VALUES	PER 100G (3.5OZ)	PER HALF PACK
ENERGY	132 kcal 552 kJ	336 kcal 1409 kJ	FAT	6.2g	15.8g
PROTEIN	6.5g	16.6g	of which SATURATES	3.3g	8.4g
CARBOHYDRATE	12.5g	31.9g	MONO-UNSATURAT	2.2g	5.6g
of which SUGARS	2.4g	6.1g	POLYUNSATURATE	0.7g	1.8g
STARCH	10.1g	25.8g	FIBRE	0.5g	1.3g

£1.59

SAINSBURY'S

HEAT AND SERVE

VEGETABLE LASAGNE

Suitable for Vegetarians

510 gram

KEEP REFRIGERATED

Use by

21 Dec

Spinach pasta layered with a selection of fresh vegetables, in a tomato sauce topped with a creamy sauce and vegetarian cheddar cheese

MICROWAVE GUIDELINES

Remove sleeve and lid. Transfer contents to a similar size and shape non-metallic dish. Cover and microwave on full power for:

Heating Category	Wattage (IEC 705)
minutes	minutes

Select appropriate time for your oven. For different power ovens, adjust the heating time accordingly.

MICROWAVE GUIDELINES FROM FROZEN

Defrost thoroughly and follow guidelines from chilled. Always check that the product is piping hot before serving.

This Package uses a minimum of 85% of recycled board

Produced in the UK for J Sainsbury plc Stamford St London SE1 9LL

00 23234 676678 76

INGREDIENTS

WATER, COOKED PASTA (DURUM WHEAT SEMOLINA, WATER, EGG, SPINACH POWDER, OLIVE OIL), TOMATOES, VEGETARIAN CHEDDAR CHEESE, COURGETTES, AUBERGINES, ONIONS, SKIMMED MILK POWDER, TOMATO PUREE, STARCH, WHEATFLOUR, MARGARINE, DIJON MUSTARD, MUSHROOMS, GREEN PEPPERS, RED PEPPERS, VEGETABLE BOUILLON, SALT, VEGETABLE OIL, BLACK PEPPER, NUTMEG, THYME, GARLIC PUREE, OREGANO, SUGAR, BASIL.

STORAGE INSTRUCTIONS

Do not reheat. If freezing, freeze on day of purchase. Consume within one month.

Copyright Ridgwell Press, PO Box 3425, London SW19 4AX. This page may be photocopied by purchasing institution only.

Making a muesli

Food activity

You can work out how a food is made by taking it apart (**disassembly**). Buy a packet of muesli and use the label to get information on the ingredients and then sort out and weigh the contents. A luxury muesli is best for this activity as it usually contains larger pieces of fruit which makes sorting easier.

Disassembly

Activity - work in groups

You need

packet of luxury muesli
accurate scales
teaspoons
large and small bowls

To do

1. Use the information from the "Ingredients" part of the label to make a **list of the ingredients** in the muesli.
2. Empty out the contents of the packet into a bowl and weigh the muesli to get the total weight.
3. Separate out the larger ingredients such as almonds and raisins and put into bowls. Tip - shake the muesli vigorously in the bowl so that the large pieces rise to the top.
4. It is difficult to sort the smaller ingredients such as wheat and oat flakes into different types. Instead find the **total weight** of these smaller ingredients and make list of the ingredients.
5. **Weigh** each type of ingredient and write down its **appearance** and its **function** in the muesli. Complete a chart like the one below. This chart is like a **product specification.** It shows the **weight** and **quality** of the ingredients in the food product.

Food product analysis			
Ingredients	**Weight**	**Appearance of ingredient**	**Functions of ingredient**
<u>Total weight =</u> *<u>Cost =</u>*			

Chart to show muesli specification

Copyright Ridgwell Press, PO Box 3425, London SW19 4AX. This page may be photocopied by purchasing institution only.

Making a muesli

Food activity

Design and make some muesli

Design and make some muesli. Use your research to help write a **recipe/formulation** for the muesli. You will not yet be able to work out the quantity of the ingredients.
The recipe could use the following headings:
* Foods for sweetness
* Foods for crunchiness
* Foods for bulk (carbohydrate)
* Foods for flavour
* Foods for texture.

Fill in this chart which shows the **function** of each ingredient.

Muesli

flavour

texture

bulk

crunchiness

sweetness

Function of foods in the muesli

Designing and making

Make and taste the muesli. Work out the quantity of each ingredient you will use as you go along. You will need to taste as you make! Now taste the finished muesli to see if you like it. Does the recipe need changing? Suggest what might be done.

Tasting

How do you think the muesli should look and taste? Draw up a list of words which could be used for tasting session when the muesli is finished.
Tasting chart

	☹	😐	☺
sweetness			
crunchiness			
flavour			
texture			
appearance			

Copyright Ridgwell Press, PO Box 3425, London SW19 4AX. This page may be photocopied by purchasing institution only.

Information from muesli

Written activity

Labels on food packets provide plenty of information about the product inside. Below are details from a Luxury Muesli packet to show **ingredients** and **nutritional information.**

Label for Fruit and nut muesli with 50% exotic fruit, nuts and seeds

Luxury Muesli Ingredients:

wheatflakes, oatflakes, raisins, dates, banana, coconut, roast hazelnuts, apricot, sunflower seeds, almonds, brazil nuts, sugar, malt extract, salt, preservative (sulphur dioxide).

Nutrition information

Typical values per 100g

energy 1572kJ (373kcal)
protein 8.5g
carbohydrate 56.3g
(of which sugars 22.5g)
fat 12.3g
(of which saturates 3.6g)
sodium less than 0.1g
fibre 9.0g

Questions

Use the label and the chart to answer the questions.

1. What is the main ingredient in the muesli on the label shown?
2. Which ingredients are added to give a) bulk, b) sweetness, c) crunchiness?
3. Which do you think is the a) most expensive b) least expensive ingredient in the muesli?
4. If you were the manufacturer and had to make a cheaper version of this muesli, what changes would you make?
5. What is the main nutrient provided by this muesli?
6. Which ingredients provide dietary fibre in this muesli recipe?

Copyright Ridgwell Press, PO Box 3425, London SW19 4AX. This page may be photocopied by purchasing institution only.

What is muesli made from? **Written activity**

Questions

The chart shows the results of disassembling a muesli. Use this chart to fill in the **Food Function** boxes in to show the use of foods in muesli.

Luxury muesli			
Ingredients	**Weight**	**Ingredient appearance**	**Function in muesli**
wheatflakes	part of 262g	flakes	bulk
oatflakes	part of 262g	flakes	bulk
raisins	104g	whole	sweetness
dates	30g	chopped bits	sweetness
banana	42g	candied slices	sweetness, crisp
coconut	8g	flakes	texture, colour
roast hazelnuts	22g	whole, skinned	crunchiness
apricot	6g	tiny dried pieces	sweetness
sunflower seeds	10g	whole seeds	flavour, crunchiness
almonds	16g	whole with skins	flavour,crunchiness
brazil nuts	part of 262g	tiny flakes	flavour, crunchiness
sugar	part of 262g	can't find	sweetness
malt extract	part of 262g	can't find	flavour
salt	part of 262g	can't find	flavour
preservative	part of 262g	can't find	preservative
Total weight 500g		**Cost £1.45**	

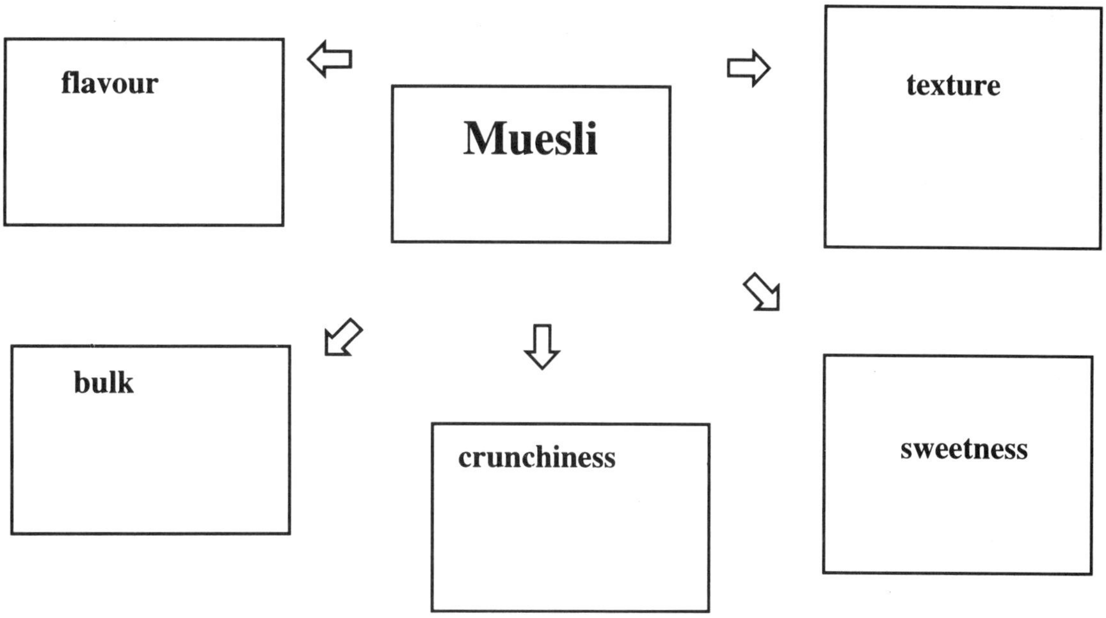

Food Function

Copyright Ridgwell Press, PO Box 3425, London SW19 4AX. This page may be photocopied by purchasing institution only.

Making a hot bread snack

Food Activity

Design brief
Design a hot bread snack.

Disassembly

Taking apart (disassembling) a hot bread snack

You can get lots of design ideas by taking food apart (disassembly). Look closely and taste a ready to eat hot-bread snack to find out the size, shape, ingredients, how it is made and what it tastes like. Use this information to help design your own hot-bread snack.

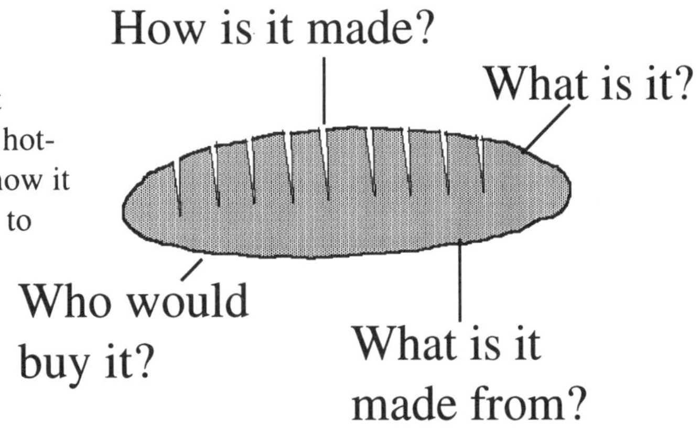

Activity

Work in groups

You need

packets of part-baked savoury or sweet bread snacks
accurate scales, knife, clean ruler, spoons

To do

1. Look closely at one of the breads before it is baked. Measure and weigh it and record this information using drawings and sketches. You can fill in one of the charts on Food Product Analysis.
2. What is the bread made from? Look at the list of ingredients for help.
3. How is the bread snack made?
3. Bake the bread according to instructions.
4. Carry out a tasting session. Describe how it looks and tastes.

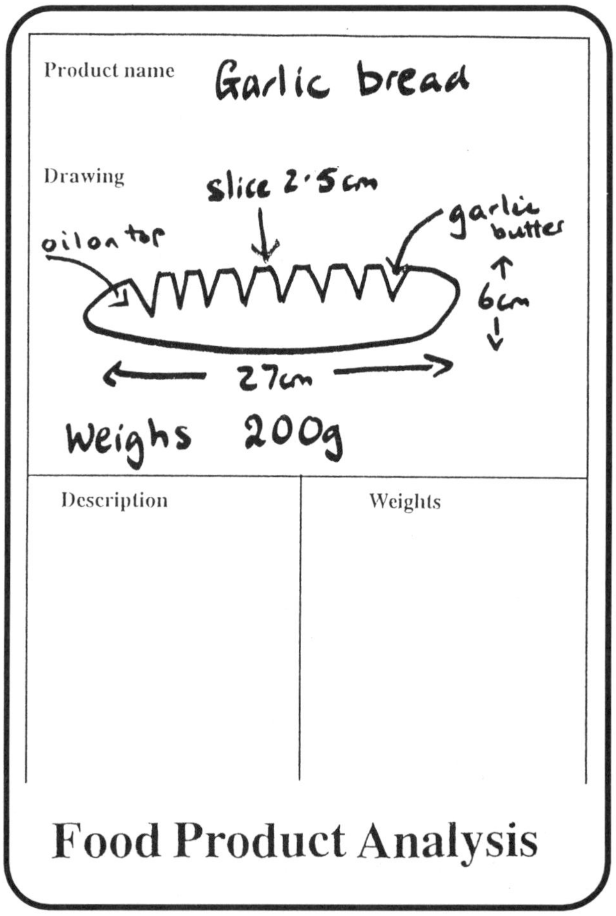

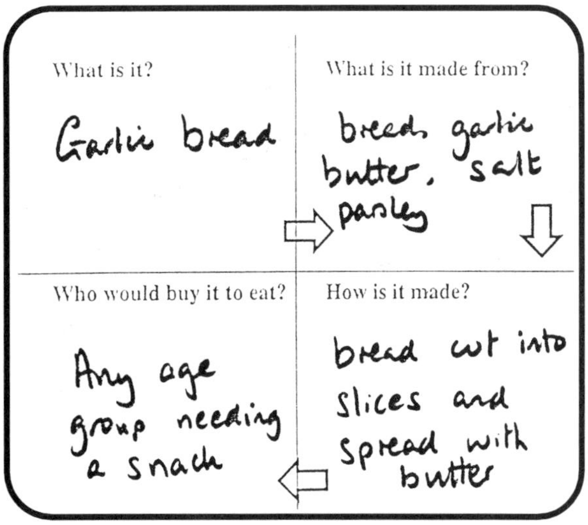

Writing the specification

Write the specification for the hot bread snack you have just looked at. Include details of the size, shape, weight, looks and taste.

Copyright Ridgwell Press, PO Box 3425, London SW19 4AX. This page may be photocopied by purchasing institution only.

Making a hot bread snack

Written activity

Design a hot bread snack
Now that you have looked at ready made hot bread snacks, use this information to design your own.
These are some of the things you need to think about:

- ❑ What type, size and weight of the bread will you use?
- ❑ What ingredients will give flavour?
- ❑ Work out the weight of the ingredients to use.
- ❑ How long will it take to cook?

Function of bread snack ingredients
Each of the ingredients in the bread snack has a purpose (**function).** In the Marmite Hot Bread Snack the function of the ingredients is:
Bread ingredient - French bread stick,
Flavouring and moistening - butter, Marmite, salt.

Design and make
Use all the research from disassembling other bread products and come up with your own ideas. Make and taste the hot bread snack and suggest improvements.

Brainstorm your ideas

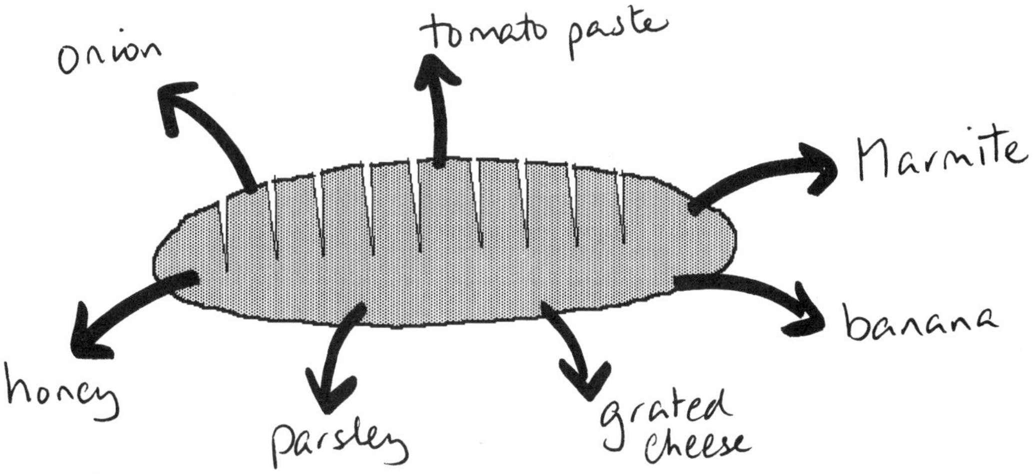

To do
List your ideas for ingredients using the headings on this chart.

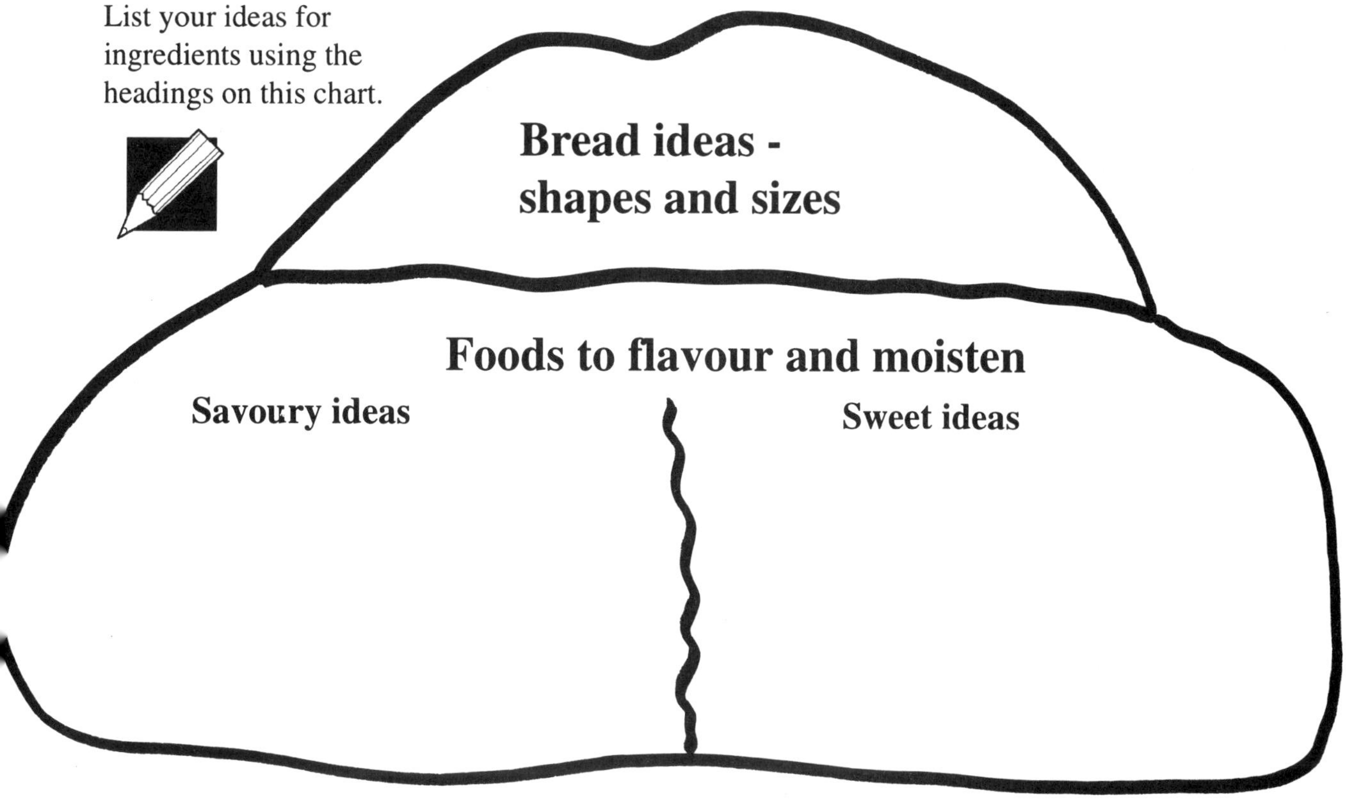

Copyright Ridgwell Press, PO Box 3425, London SW19 4AX. This page may be photocopied by purchasing institution only.

Bread quality check

Written Activity

Choosing for quality
Choose words you want to describe the quality of your food product.

Quality checklist
I want my garlic bread to be:

safe and clean to eat	☑
crisp and evenly cooked	☑
hot	☑
tasty	☑

You can write the method (**making process**) to show how you will build **quality checks** into the way you make garlic bread (or a hot bread snack).

Quality Assurance
Taste the bread when it is cooked. If it doesn't taste and look as you expected, look through your **quality checklist** and see where you should have made changes.

How will you get this standard of quality?
How you will get this quality as you make the product?

⇩

Quality words	Quality checks needed
Safe and clean to eat	*Clean up before you start.*
Crisp, hot and evenly cooked	*Bake in the centre of hot oven for correct time. Test with a temperature probe.*
Tasty and attractive	*Use enough garlic and salt and don't over or undercook.*

Garlic bread recipe
French bread stick, 100g butter, 3 cloves crushed garlic, chopped parsley
Set the oven at 200°C/Gas 6. Cut the loaf into slices. Beat together the butter, garlic and parsley. Spread the mixture between the slices of bread. Wrap the loaf in foil and bake 15 minutes.

To do
Draw a Quality Checklist chart like the one below. Use the recipe for garlic bread and fill in the chart to show the Making process and Quality Checklist. One step has already been completed.

	Making process	Quality Checklist
1	Collect bread, butter, garlic and parsley	Make sure bread is fresh.

Copyright Ridgwell Press, PO Box 3425, London SW19 4AX. This page may be photocopied by purchasing institution only.

Bread quality check

Written Activity

This is a simple list of some **quality checks** a food factory would need when making garlic bread.

Chart showing quality checks for ready-made garlic bread in a food factory

Factory Process	**Quality Checklist**	**What to do if quality not met**
Choose ingredients	Are the ingredients the right quality and weight?	Go back to supplier and change
Prepare the garlic bread	Is the process clean and safe?	Clean up factory and close if necessary.
Bake the bread	Is the bread properly baked?	Reject burnt and undercooked breads.
Fill the cooled bread with garlic butter	Make sure the bread is cut and filled evenly with butter.	Reject breads which are not properly filled.
Pack for sale.	Is the packaging safe and instructions clear?	Reject poor quality packaging.
Customer buys and bakes garlic bread.	Customer decides if bread is properly cooked, tastes nice and is safe to eat.	Take back to shop if it does not meet quality standards.

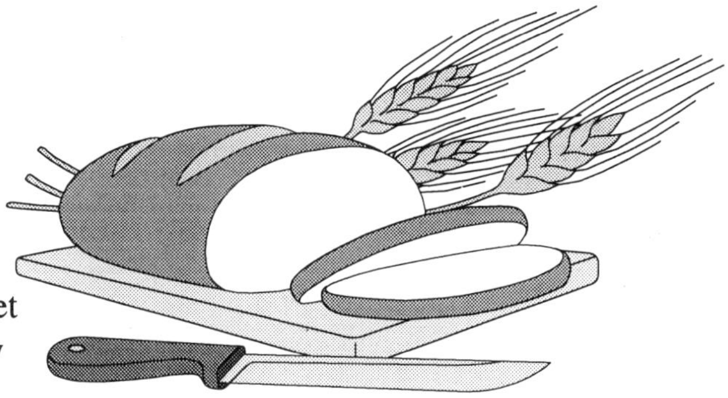

Quality Assurance
A food factory must check food products regularly to see if they meet quality standards. If not, the quality control systems need changing.

Questions

1. What went wrong if the garlic bread was:
1) burnt at one edge, 2) burnt all over, 3) not very tasty, 4) not hot?
2. What would need to be changed in the way the bread was made to improve each of these problems?
3. Compare the charts showing the quality checklists for a) home-made garlic bread, b) ready-made garlic bread. What do you think is different between the way garlic bread is made at home and in a food factory?
4. How do quality checks help food manufacturers to make large quantities of food products?

Copyright Ridgwell Press, PO Box 3425, London SW19 4AX. This page may be photocopied by purchasing institution only.

Hot bread snack labels

Written activity

What information comes from bread labels?

This chart shows the ingredients from the labels of 3 different hot bread snacks: **Marmite Bread, Cheese Bread and Garlic Bread.**

Ingredients	Marmite bread	Cheese bread	Garlic bread
wheatflour	*	*	*
butter	*	*	*
water	*	*	*
yeast	*	*	*
flour improver	*	*	*
ascorbic acid	*	*	*
salt	*	*	*
dextrose	*		
Marmite	*		
cheese		*	
cream cheese		*	
gluten		*	
cheese flavouring		*	
garlic puree			*
parsley			*
lemon juice			*

Ingredients chart

Questions

1. Use the chart to make a list of the ingredients used to **flavour** each of these breads. To find out this information, look at the ingredients that are the **same** for each bread. The other ingredients are used for flavouring.

Ingredients used to flavour breads:

Marmite bread

____

____

Cheese bread

____

____

Garlic bread

____

____

2. List three ingredients on the Ingredients chart which might be difficult to buy in the shops. What do you think these ingredients are used for?

3. What ingredients do you think are used to make the bread?

Copyright Ridgwell Press, PO Box 3425, London SW19 4AX. This page may be photocopied by purchasing institution only.

Hot bread snack labels

Written activity

Information from bread labels

This chart shows the nutrition information from the labels of 3 different hot bread snacks: **Marmite Bread, Cheese Bread and Garlic Bread.**

Nutrition Information chart

Nutritional value in 100g	Marmite bread	Cheese bread	Garlic bread
Energy	*1605 kJ/384 kcal*	*1498 kJ/358 kcal*	*1591 kJ/ 379 kcal*
Protein	*9.1g*	*10.1g*	*7.9g*
Carbohydrate	*39.6g*	*39.7g*	*49.8g*
of which sugars	*2.2g*	*1.9g*	*1.9g*
Fat	*21.0g*	*17.6g*	*16.5g*
of which saturates	*12.3g*	*11.8g*	*10.5g*
Fibre	*1.7g*	*3.3g*	*3.9g*
Sodium	*0.7g*	*0.6g*	*0.6g*

Questions

Use the Nutritional Information chart for the 3 breads to answer the questions.

1. Which bread is the most "fattening" - has the highest energy value (in KJ/ kcals)?
2. Why do you think Cheese bread provides the most protein in 100g portion?
3. Which bread has the highest amount of fat in 100g? Which ingredients provide this fat?
4. Which ingredients provide fibre and sodium in these breads?

Copyright Ridgwell Press, PO Box 3425, London SW19 4AX. This page may be photocopied by purchasing institution only.

Making a fruit pie

Food activity

What is a fruit pie made from and how is it made? Carry out this investigation to take apart a ready-to-eat fruit pie. This is know as **"disassembling a product"** and is used to give design ideas.

Investigation

You need

Some ready-to-eat fruit pies with packaging
Teaspoons and plates for tasting, accurate scales and a clean ruler.

To do

❑ Look carefully at how the pies are packed. Find out the total cost of the pie or pies and if there are several pies in a packet, work out the cost of each pie.
❑ Draw and label the pies to show how they are made.
❑ Take apart a pie - weigh the pastry and the filling. Record your results.
❑ Taste the pie and describe how it looks and tastes.
❑ Make a note of the list of ingredients and the nutritional value of the product.
❑ You could fill in some charts likes the ones shown for Food Product Analysis.

Writing a Specification
Write a **specification** for the pie which shows details of the size, shape and decoration. Describe how you think the pie is made. Add words to describe taste and quality.

Food Product Analysis

Product name
Lyons fruit tarts

Drawing

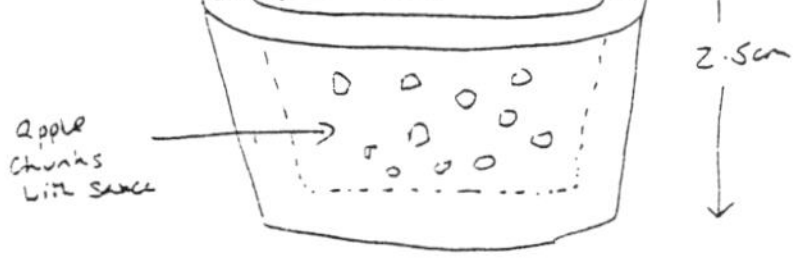

Description	Weights
Light gold pastry tart with round top sprinkled with sugar. *Filling - cubes of chewy dried apple with pale yellow gluey sauce.*	*Weight of fooc* *54g* *Weight of foi* *Weight of pa* *Weight of fil*

Food Product Analysis

Name of the food product
Mr Kipling 6 Fruit Pies Selection

Food product description/ Best before/ Cost per pie
2 apricot, 2 apple and blackcurrant, 2 Bramley apple
No artificial colours. Baked with vegetable fat.
Suitable for vegetarians.
Best before 14 Oct (purchase 27 Sept) Cost per pie 14p

Statements showing product quality
These products have been carefully made from the finest ingredients and we want you to enjoy them at their best.

Ingredients in order their weight and percentages
Wheatflour, sugar, hydrogenated vegetable fat, apple (10%), apricots (7%), glucose syrup, blackcurrants (3%), dextrose, modified starch, gelling agents: pectin, sodium alginate; salt, citric acid, preservative: potassium sorbate; flavourings, adipic acid, colours: lutein, anthocyanins; acidity regulator: sodium citrate.

Any other information on quality
"I'm told that my 'Thumbprint pies are causing quite a dilemma," said Mr Kipling. "The choice between my scrumptious Bramley Apple Pies and my Apple and Blackcurrant Pies is proving to be a considerably difficult one." "Never mind," exclaimed Mr Kipling, I've decided to introduce a selection of my favourites including tangy Apricot Pies - an exceedingly good choice for all the family!"

Copyright Ridgwell Press, PO Box 3425, London SW19 4AX. This page may be photocopied by purchasing institution only.

Making a fruit pie

Design brief
Design and make a delicious fruit pie to serve 4.

Write a specification for a fruit pie

Now you have taken apart a ready-to-eat pie, use the results of the investigation to help you write a **specification** for your own fruit pie to fit the design brief.

A specification can be like a recipe. You can make a **list of ingredients** and the **amounts** that that you will use. Describe how the pie will **look** and **taste.** You need details of **size** and **shape**. You may give the **cost** and **nutritional information** and how the pie should be **stored** and how long it will **keep fresh**.

To sum up

A specification for a food product can provide information on:

- ❐ **size**
- ❐ **shape**
- ❐ **ingredients**
- ❐ **cost**
- ❐ **product quality**
- ❐ **nutritional value**
- ❐ **storage time**

You could draw and label the type of pie you want to make to show the specification.

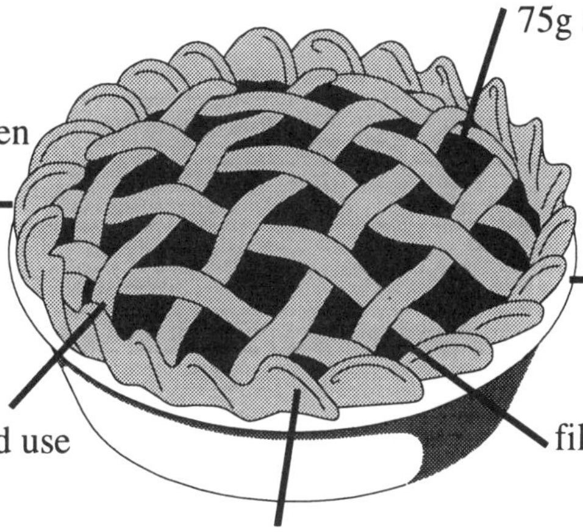

A written specification:

Apple Pie Specification
20cm round pie, serves 4
Ingredients: *150 g shortcrust pastry* - 100g flour, 50g butter, 1 tablespoon water, *filling* - 500g Bramley apples, peeled and sliced, 75g brown sugar, cost about £1, crisp pastry with tangy filling.

Question

Write a brief specification for one of the following foods: a) a fruit drink, b) a biscuit, c) a bag of crisps. You don't need to include details of an exact recipe.

Copyright Ridgwell Press, PO Box 3425, London SW19 4AX. This page may be photocopied by purchasing institution only.

Fruit pie labels

Written activity

Food manufacturers spend thousands of pounds researching and developing the recipe for a food product, so the details and product specifications are usually a company secret. But you can get plenty of information from a label. Use the information from these three labels for fruit pies to answer the questions on page 21. When you have finished, discuss how much information you have found.

Nutrition Information

Sainsburys 6 Assorted Fruit Pies

2 apple, 2 blackcurrant and apple, 2 strawberry

Cost 9p per pie.

Ingredients

Wheatflour, sugar, hydrogenated vegetable oil, apples, glucose syrup, strawberries, blackcurrants, modified starch, dextrose, salt, stabiliser: sodium alginate; citric acid, preservative: potassium sorbate; adipic acid, flavouring, vegetable protein

Sainsbury's	assorted pies	
	per 100g (3.5oz)	per pie
Energy	349 k Calories 1469 k Joules	187 k Calories 786 k Joules
Protein	3.2g	1.7g
Carbohydrates	60g	32g
of which sugars	30.7g	16.4g
starch	29.1g	15.6g
Fat	10.7g	5.7g
of which saturates	2.2g	1.2g
monounsaturates	6.4g	3.4g
polyunsaturates	2.0g	1.1g
fibre	1.4g	0.7g
sodium	0.1g	0.006g

Lyons Assorted Fruit Pies

Apple pies, raspberry and redcurrant, blackcurrant and apple

Cost 11p per pie

Ingredients

Wheatflour, sugar, hydrogenated vegetable oil, apple (9.7%), dextrose, raspberries (3.6%), redcurrants (3.0%), blackcurrants (2.8%), modified starch, starch, citric acid, malic acid, salt, preservative:potassium sorbate; flavouring.

Lyons Fruit Pies	Nutrition	Information
Typical values	per 100g	per pie
Energy	389 kcal 1635 kJ	204 kcal 861 kJ
protein	3g	1.6g
carbohydrate	59.7g	31.4g
fat	15.3g	8.1g

Mr Kipling 6 Fruit Pies Selection

2 apricot, 2 apple and blackcurrant, 2 Bramley apple

Cost 14p per pie

Ingredients

Wheatflour, sugar, hydrogenated vegetable fat, apple (10%), apricots (7%), glucose syrup, blackcurrants (3%), dextrose, modified starch, gelling agents: pectin, sodium alginate; salt, citric acid, preservative:potassium sorbate; flavourings, adipic acid, colours: lutein, anthocyanins; acidity regulator: sodium citrate.

Mr Kipling Fruit Pies	Nutrition	Information
Typical values	per 100g	per pie
Energy	359 kcal 1510kJ	241 kcal 1015 kJ
Protein	3.2g	2.1g
Carbohydrate	58.4g	39.2g
Fat	12.5g	8.4g

Copyright Ridgwell Press, PO Box 3425, London SW19 4AX. This page may be photocopied by purchasing institution only.

Fruit pie labels

Written activity

Questions

Nutritional information

Use the "Nutritional Information" section on each label for these questions.

1. Which label gives the most detailed nutritional information? How can customers use nutrition information on a label?
2. Compare the amount of nutrients in each of the three pies. You could draw up a chart and show the amount of energy, protein, carbohydrate and fat in 100g of each of the pies.
Which pie provides the most energy (kJoules/kCalories)? Explain why you think this pie has the highest energy value.
3. One label shows the amount of sugar in each pie. How many teaspoons of sugar does this represent?

Did you know?
The new proposed system on food labels (QUID) will show the percentage of the main ingredients in the food product - for example, the fruit pie labels show fruit content.

Questions

Ingredients

Use the "Ingredients" listed for each pie to answer these questions.

1. Choose one of the pies. List the ingredients which you think are used for: a) the pastry, b) the filling in the pies.
2. Two labels show the percentage of fruit in the pies. Why do you think this information is useful?
3. What ingredients are used for a) sweetening the pies, b) fats?
4. List the following types of additives used in the pies: a) preservatives, b) stabilisers, c) gelling agents, d) colours, e) acidity regulators.
5. Give three reasons why it would be difficult to make fruit pies like the ones in these packets.

Idea - using a computer, create a database to show how different additives are used in food products. Use food labels to collect this information. This information is helpful for food technology studies. You could set up the fields shown on the chart below.

type of additive	examples	found in..
preservatives	potassium sorbate	apple pies

Answers to questions on page 51.

Copyright Ridgwell Press, PO Box 3425, London SW19 4AX. This page may be photocopied by purchasing institution only.

Quality of food

Design brief - Design and make a delicious fruit pie to serve 4.

Quality Checklist for apple pie

hygienically prepared	☑
safe to eat	☑
looks attractive	☑
thoroughly cooked	☑
tastes good	☑
crisp pastry	☑
serves 4 people	☑
skillfully finished	☑

Quality control
In order to control the quality of a product, you need a **quality checklist** to describe the standards of quality you want to meet.

Quality control and Planning
Plan how you will make the pie. In the plan, show how you will try and control the quality as you work. You can tick off these quality controls as you make the pie. An example of quality control is shown on the timeplan.

Make the pie
When the food is ready to eat, you can check to see if you have met the standards set down on the **quality checklist**. Carry out a taste testing to find out how it looks and tastes. Finally, check that the pie meets the needs of your **specification.**
If it does, you have designed and made a pie which meets your **design brief.**

Time	Order of Work	Quality Control
9.25	Set oven 200°C / Gas 6	
9.30	Make pastry by rubbing in method	Make sure it is not too sticky or wet.
9.40	Roll out pastry and line dish.	Don't over roll pastry.
9.45	Peel and slice apples and sprinkle on sugar.	Clear away peelings. Slice apple evenly.
9.55	Roll out pastry strips. Place neatly on pie.	Make sure edges and strips are even.
10.05	Bake in centre of oven for 30 minutes.	Check baking is even.
10.30	Remove from oven.	Is it properly cooked?

Copyright Ridgwell Press, PO Box 3425, London SW19 4AX. This page may be photocopied by purchasing institution only.

Quality work

Food activity

How can you show others that you have tried to make a **quality food product**? A photograph of a food product may show the quality of the finished food product, but a photograph can fib - the food may not be properly cooked or taste very good.
The way you work and the **quality control checks** that are built into the way you work can lead to a good quality food product.
You can use this **Quality Report** to show how during planning and making you may or may not end up with a well made food product which is safe and tasty to eat.

Quality Report -Tick a box ☑ for each statement

Quality Report

	Yes	Needs improving	No
Worked safely	☐	☐	☐
Worked hygienically	☐	☐	☐
Stored and cooked food at safe temperatures	☐	☐	☐
Used knowledge and skill	☐	☐	☐
Worked to plan	☐	☐	☐
Worked efficiently	☐	☐	☐
Cleared up	☐	☐	☐
Presented work on time	☐	☐	☐
Made a quality, tasty food product	☐	☐	☐

What's the difference between...?
Quality of design - The design team uses skills and knowledge to create a product and explains how it should be made.
Quality of manufacture - This is the way the product has been made according to the instructions of the design team.

A design may be good, but the way it is made may not be. For example, a supermarket makes a very good ready-to-eat lasagne, but if you burn it, the end product is no good.
To do - Think of your own examples of products which might be well designed, but could be poorly made.

Children's food

Food activity

Task
Design and make a food product for an under 10 year old.

First, investigate what is on sale for children in supermarkets and restaurants then disassemble some food products.

Disassembly

Work in groups.
Collect a range of children's food products to disassemble.
Choose one product to disassemble in your group.
You can use some forms in this book to help.

Food Product Analysis Taking apart (disassembling) a food product to get design information Name of food product
Boots Teddy Bear biscuits for toddlers
How does it look and taste?
8 bear shaped biscuits, golden brown colour, very sweet and crunchy
How is it made?
Dough is mixed and rolled out. Dough is cut into bear shapes, then baked.
Other useful information
Made with honey and added vitamins and minerals. Food to provide toddlers with fun, taste and nutrition.
What ideas does it give you for food product design?
Children like food in fun shapes. Packaging and healthy food are important.

Design criteria

Use the results of the disassembly exercise to come up with **design criteria** for a food product for young children.
Think about:

- ❑ what children need to eat
- ❑ what food they like to eat
- ❑ what makes the food appeal to them
- ❑ other things you think are important.

As you disassemble, make a note of:

The way the packaging is made and designed - draw the details.
Look at the size and shape of the product.
How is it made?
What is it made from?
Why is it suitable for young children?
Describe the looks and taste of the product.
Write a specification for the food product.

Questions

1. Make a list of 10 points of important things to consider when designing children's food.
2. Draw a star profile to show eight of these points which you would use in your design and give them a rating of 0-5, where 0="not.." and 5= "very.."

Copyright Ridgwell Press, PO Box 3425, London SW19 4AX. This page may be photocopied by purchasing institution only.

Children's food

Food activity

> **Design and make**
> Design and make ideas for a food product suitable for an under 10 year old.

Make a **Checklist** or **Design star profile** showing the design criteria you will be working to.
Design star profile

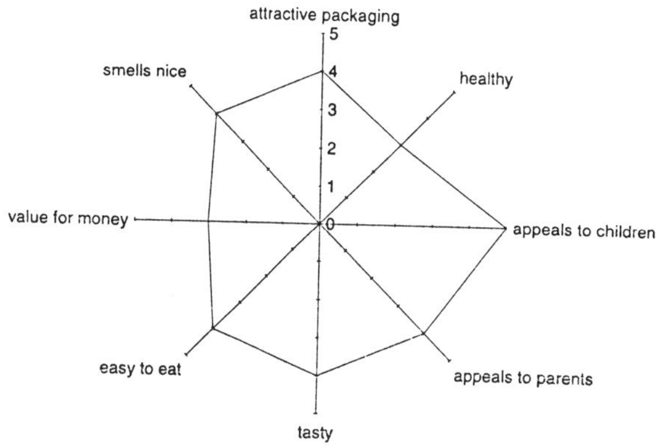

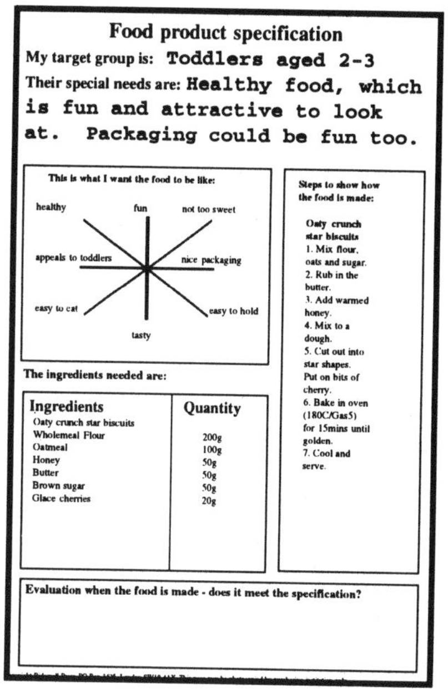

Food product specification

My target group is: **Toddlers aged 2-3**

Their special needs are: **Healthy food, which is fun and attractive to look at. Packaging could be fun too.**

This is what I want the food to be like:

healthy, fun, not too sweet, appeals to toddlers, nice packaging, easy to eat, easy to hold, tasty

The ingredients needed are:

Ingredients	Quantity
Oaty crunch star biscuits	
Wholemeal Flour	200g
Oatmeal	100g
Honey	50g
Butter	50g
Brown sugar	50g
Glace cherries	20g

Steps to show how the food is made:

Oaty crunch star biscuits
1. Mix flour, oats and sugar.
2. Rub in the butter.
3. Add warmed honey.
4. Mix to a dough.
5. Cut out into star shapes. Put on bits of cherry.
6. Bake in oven (180C/Gas5) for 15mins until golden.
7. Cool and serve.

Evaluation when the food is made - does it meet the specification?

When you have a clear idea of what you want to make, you could complete a **Food product specification chart** like the one shown.

This star profile for children's food shows the design criteria which one group will use in their design work.

During the planning stage, you can build in steps for **quality control** as you make the product. Use one of the charts on page 49.

Now make your food and taste and test it to see if it meets the standards you have set in the **Checklist, Star profile or Specification.** You could ask young children to be on your tasting panel - they will tell you if your product suits them! Make any changes they suggest.

Plan a tasting session

Questions

Each year there are more and more food products for young children. Think of 4 food products and give the reasons why you think young children like them.

Food product	Why do young children like it?
1.	
2.	
3.	
4.	

Copyright Ridgwell Press, PO Box 3425, London SW19 4AX. This page may be photocopied by purchasing institution only.

Children's food

Written activity

Pester power - food and drink with child appeal

The children's food and drink market is growing in size and importance for food manufacturers. In 1993, the total value of the UK children's food and drink market was valued at £6 billion. Children are becoming more independent and have more spending power. They influence family purchases. 15% of parents of 6-12 year olds claim that their children exert a great influence on family shopping. Children's favourite foods include savoury snacks such as crisps, then sweets and drinks.

What is the food industry doing about children's foods? **Character merchandising** is being used to increase child appeal - this means using characters such as The Flintstones to promote popcorn, and Mr Blobby to sell cakes and sweets. Some food producers create their own characters such as the "Munch Bunch" yogurt range.

If a product is successful in one area, it may be developed into a new type of product. Mars bars have been made into icecream and drinks. Many food companies have research and development teams that closely watch trends to predict new food fads.

Children's and adult's taste

There are major differences between children's and adult's taste. Taste buds number around 10,000 in small children and around 2,000 in adults. However, children's taste buds are untrained but adults have experienced many foods and know what they like and dislike. Children like basic flavourings such as orange, vanilla, chocolate and lemon. Children like the way a food looks and they are more willing to try new products if they look good.

Questions

1. What special foods do you ask your family to buy for you? Give your reasons for these choices.
2. Name 6 new food products which have been developed for children in recent years.
3. Give 3 examples of how "character merchandising" is used for children's foods.
4. Design the packaging for a children's food and explain why it appeals to children.

Copyright Ridgwell Press, PO Box 3425, London SW19 4AX. This page may be photocopied by purchasing institution only.

Children's food

What do you think about this ready meal for toddlers? Read the label to help answer the questions.

Chicken Hot-Pot

Healthy ready-meal suitable for toddlers from one year

This delicious, nutritious meal for toddlers has been specially chosen for young children. It is made with small pieces for little mouths and has added vitamins and minerals.

You can eat the meal in the tray and use the carton as a fun way for your toddler to eat its meal.

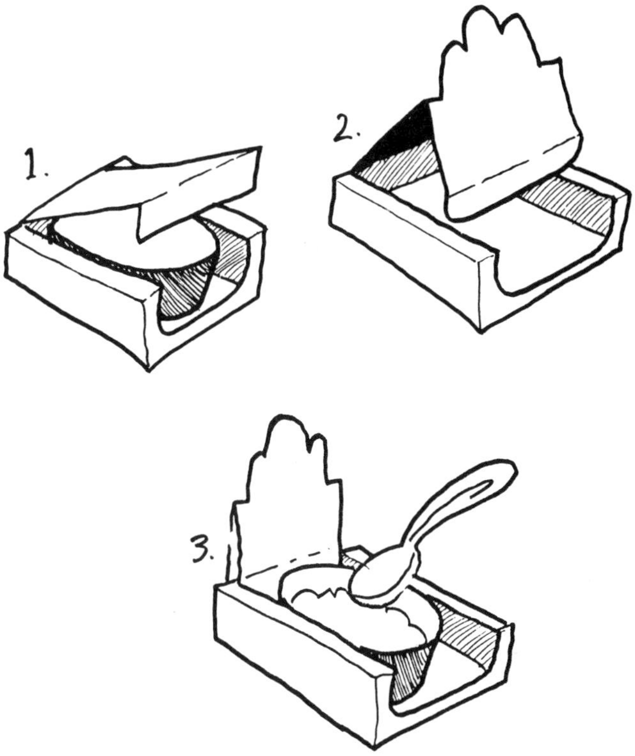

CONVENTIONAL COOKING: Remove the outer packaging and put the sealed tray in a saucepan of boiling water. Boil and heat for 15minutes.

MICROWAVE COOKING: Pierce the film lid in several places with a sharp knife. Microwave on medium high (70%) power for 2 ½ minutes (650 watts). Stand for 30 seconds. Carefully remove the lid and serve.

IMPORTANT:Check the temperature of the product before serving. Once heated this product should not be reheated. Do not leave your child unattended with any food.

Questions

1. Evaluate the Chicken Hot-Pot dish. Do you think this product will appeal to a toddler?
2. What do you think about the packaging and the way this product could be served in the carton?
3. Do you think there is a need for this type of product? Give your reasons.
4. What are the advantages of this kind of packaged, convenience food for:

a) The manufacturer

b) The parent?

Further Work

Design an interesting or amusing package for a toddler's ready meal. Use cardboard to make a model of your design.

Copyright Ridgwell Press, PO Box 3425, London SW19 4AX. This page may be photocopied by purchasing institution only.

Quality of food

Written activity

"The history and evolution of quality control are linked with the technological advances of the human race." (Principles of Quality Control by Jerry Banks).

What is quality?

"Quality is when customers keep coming back but products don't"

The success of a company depends on the quality of its products. If customers are happy with the product quality they will continue to buy. All companies need a system to make sure their products keep their standards of quality.

Quality Assurance

Quality Assurance is a system where everyone in a company takes responsibility for quality at all levels. Many larger companies have a **Quality Assurance Department** which sets standards for quality on all products.

This is the definition of Quality Assurance from BS 4778: " *Quality Assurance is all those planned and systematic actions that a product or service will satisfy given requirements for quality.*"

This is another definition:

"Quality Assurance is a system which aims to ensure that quality is inbuilt by means of written procedures and communications."

Quality Control

This is the BS 4778 definition for Quality Control:

"Operational techniques and activities that are used to fulfil requirements for quality".

This is another definition:

"Quality Control is control by inspection and a means of picking out faulty goods."

British Standards

Many thousands of British Standards exist which set down the minimum safety requirements for products and product performance and reliability. British Standards can be used to test products. Tasting is one of the most important tests for food products.

BS 5750

BS (British Standards) 5750 relates to Quality Management Systems and UK companies can gain accreditation if they reach high enough quality standards. The companies must show documentary evidence for quality procedures.

Hazard Analysis

Food safety controls

Food must be safe to eat otherwise people will get sick. A food business has to work out a food safety system to make sure that its food is safe for customers to eat. This **food safety system** is planned step by step by identifying the hazards at each stage and planning how to control that hazard.

A **hazard** is anything that may cause harm to the customer.

A **hazard** may be:

* biological such as salmonella in chicken
* physical such as glass in food
* chemical such as cleaning chemicals in food.

The food industry uses a system called **Hazard Analysis and Critical Control Point (HACCP)** to control food safety problems.

Questions

1. Why do you think quality of food products is important to a food business?
2. What is a) Quality Assurance, b) Quality control, c) a food hazard?
3. What are British Standards?
4. What is meant by HACCP?

Copyright Ridgwell Press, PO Box 3425, London SW19 4AX. This page may be photocopied by purchasing institution only.

Quality of food

Food activity

Food products are made to a **specification** which describes the **quality of the ingredients** used. Manufacturers check that their **quality control** is working by taking samples of the product and carefully weighing, measuring and describing the contents. Decide for yourself what the quality specification is for a food product.

Activity

Choose one of these products to investigate:
canned fruit such as peaches or apricots
canned vegetables such as carrots or beans

Work in groups

You need

canned fruit or vegetables
scales, measuring jug, bowl, spoon, sieve, clean ruler, paper towel

To do

1. Open the can and drain the contents through a sieve. Collect the liquid in a bowl. Weigh the liquid and then the fruit or vegetable.
2. Look carefully at the fruits or vegetables. Count and measure them. What is the quality? You can rate the quality on a 1-3 score: 1 = good, 2 = average and 3 = poor. Fill in the chart with you results.

Use the results to write a **specification** which gives details of size, shape, weight, number and quality of contents.

Name of canned food		
weight of can and contents	______	
empty can weight	______	
weight of liquid contents	______	
weight of fruit or vegetables	______	
number of fruit or vegetables in can	______	
size of fruit in can	______	
quality of fruit (1-3 rating)	______	
	Yes	No
whole or pieces?		
does size vary?		
does colour vary?		
other comments ______		

Food Product Specification

Size ....

Shape ....

Weight ....

Looks ....

Taste ....

Further work

1. Compare one food product with another similar product - for example, own labels from supermarkets or cheaper brands.

2. Compare the quality of ready packed fruit or vegetables such as potatoes, apples, carrots or onions. Find out number in pack, average size/ weight of fruit or vegetables and their quality.

Copyright Ridgwell Press, PO Box 3425, London SW19 4AX. This page may be photocopied by purchasing institution only.

Quality of food

Written activity

Food needs to be checked during making and cooking to make sure it is safe and good to eat. These checks are some of the **quality controls** during food production.

Quality in biscuit making

A biscuit factory needs to make constant checks to make sure that the biscuits are all the same quality. The chart for **Large scale biscuit making** shows these quality controls. You can use this skill of quality control to improve the way you make food products.

Large scale biscuit making in a factory

Process		Quality controls
Select and weigh ingredients.	⇨	Make sure ingredients are good quality and weigh accurately.
Mix ingredients to form a dough.	⇨	Dough must be right consistency for biscuit.
Knead and roll dough.	⇨	Don't overwork dough as it affects texture.
Cut dough into shapes.	⇨	Make sure shapes are same size and weight.
Bake biscuits.	⇨	Biscuits should be same colour and evenly baked.
Cool biscuits.		
QUALITY CHECK	⇨	Check appearance, weight, moisture and colour.
Pack biscuits Store and distribute.	⇨	Check and remove broken biscuits and faulty packaging. Check date stamp, stock control and storage.

Wheatmeal cookie recipe

Ingredients

225g wholemeal flour
150g butter or margarine
1 beaten egg size 3
25g caster sugar

Method

1. Set the oven at 180°C/ Gas Mark 4.
2. Put the flour and sugar in a bowl and rub in the butter or margarine.
3. Stir in the beaten egg to make a stiff dough - add water if needed.
4. Roll out the dough on a floured surface and cut into shapes.
5. Place the cookies on a greased baking tray and cook for 15-20 minutes.
6. Cool on a wire tray.

To do

Try out this exercise - use the home-made recipe for Wheatmeal cookies. Write out the **steps** in the recipe. For each step, show how you can control the **quality** to make sure the biscuits are well cooked. Use the chart for **Large scale biscuit making** to help.

Further work Add 2 extra steps to the recipe which you think will improve quality control further. Try making the cookies to show if your Quality Checklist works!

STEPS	QUALITY CHECK
1. Set oven	Must reach right temp.
2.	

Copyright Ridgwell Press, PO Box 3425, London SW19 4AX. This page may be photocopied by purchasing institution only.

Quality of food

Written activity

Large scale production of your Wheatmeal cookies (Mini enterprise)
You have a team of 4 people to help make 200 of your wheatmeal cookies to sell to raise money. You need to speed up the production and share the tasks. What jobs could each of the 4 people do to help you make 200 biscuits and clear up in less than one hour? Fill in the chart for **Large scale production** to show the jobs each person can do and the quality checks needed. Try it out to raise funds!

Large scale production chart

Person	Jobs	Quality checks needed
Tex 2	Mixing the dough	Get right consistency

Person	Jobs	Quality checks needed
John 1	Weighi[illegible] Packi[illegible]	[illegible]

Person	Jobs	Quality checks needed
Paolo 4	Baking Clearing up	

Person	Jobs	Quality checks needed
Freda 3	Roll out biscuits and cut out	Right thickness and size

Sharing the tasks

Large scale biscuit production

Questions

1. How can you change your cookies to make them appeal to sell to the following people:
a) toddlers , b) health conscious teenagers,
c) a gift for an elderly person?
2. Design and draw a label which you can attach to the packet to sell to one of these groups.

Copyright Ridgwell Press, PO Box 3425, London SW19 4AX. This page may be photocopied by purchasing institution only.

Quick questions

Written activity

1. What is meant by "disassembly"?

2. Name 4 things you can find out if you "disassemble" a food product.

3. List 4 words which you think are important to describe the quality of food.

4. List 4 things you would check to make sure that a fruit pie is good quality.

5. Give 3 reasons why food companies need to make good quality food products.

6. List 4 different types of information you can get from a food label, for example - how long it will keep.

7. List 3-5 things which could be used for the specification of a food product.

8. Draw a star profile for a fruit pie and use words on each of the lines to describe how you think the pie should look and taste. Now draw in the star using the rating of 0=not and 5=very.

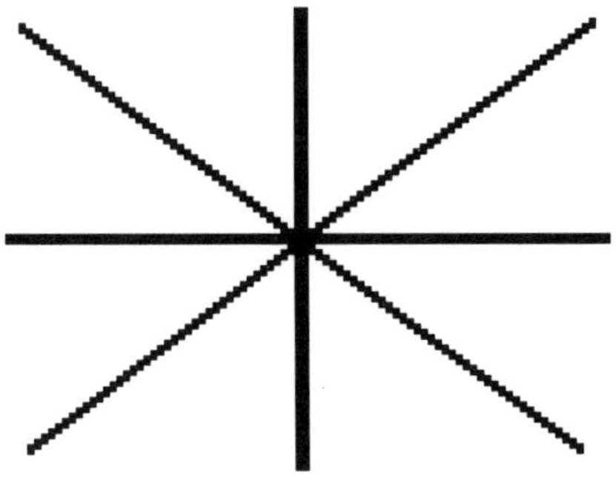

Copyright Ridgwell Press, PO Box 3425, London SW19 4AX. This page may be photocopied by purchasing institution only.

Did you know?

Written activity

Bananas must not bend

From January 1st 1995, the European Commission has new quality standards for bananas. They must not bend "abnormally" and must be a minimum of 5.5 inches long and 1.06 inches round. Buyers from the big supermarket chains want to be sure they are getting the same quality of bananas when they order huge quantities over the phone. Signs may go up in the shops saying "Yes, we have no bent bananas"!

United Nations plans the menu

In 1994, as a three month experiment, the United Nations Secretary General asked that UN troops have a similar menu because soldiers from different countries were eating different meals. A list was made to describe the exact food quality required.

Carrots must be one and a half inches in diameter, sprouts must be no less than one inch in diameter and well coloured, not withered and free from disease, and lemons must satisfy 20 criteria.

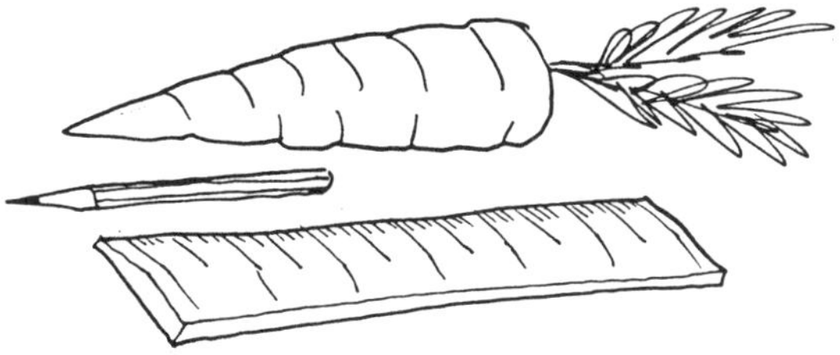

Sandwich shop taken to court for following old recipe

In 1994 and sandwich shop poisoned 62 people by serving them sandwiches containing mayonnaise made to an old recipe. The sandwich shop owner thought it was safe to make Mrs Beeton's Victorian recipe which used raw eggs. Unfortunately today's eggs are sometimes contaminated with salmonella, bacteria which cause food poisoning. Current advice from the Government states that eggs should not be eaten raw. The sandwich shop owner was fined and advised to use pasteurised mayonnaise in future.

Questions

1. Why do you think quality control is needed for foods such as bananas?
2. What do you think about all the United Nations troops eating the same menu?
3. Lemons can have 20 criteria for quality. Come up with 5 words to describe the quality of a lemon.
4. Why might raw eggs be dangerous to eat?

Copyright Ridgwell Press, PO Box 3425, London SW19 4AX. This page may be photocopied by purchasing institution only.

Teacher's notes

This book focuses on Food Activities which **disassemble food products** and looks at the importance of **quality control in food work.**

What is disassembly?
Disassembly means **"taking apart"** and is a powerful tool for understanding design of any kind. The food industry frequently examines other products to get ideas and information by disassembly, and this process is mirrored by activities in this book.
In the book, practical activities and written activities are shown with symbols:

for practical activities

for written activities

How does this book address the National Curriculum Orders?
These statements are taken from the Orders:
Pupils should develop design and capability through:
b Focused practical tasks in which they develop and practise particular skills and knowledge.
c Activities in which they investigate, disassemble and evaluate familiar products and applications.
This book focuses on both of these sections.
On the next page, statements from the orders which are arrowed and in bold are areas covered by ideas in this book.
Focused practical tasks
There are four sections with focused tasks:
1.Design and make a muesli, 2. hot bread snacks, 3. fruit pies and 4. children's food.
The aim of the activities is to create short tasks which build on the knowledge and skills needed to **disassemble food products** and understand the importance of quality control. Each section - muesli, hot bread snack, fruit pies and children's food - is presented as a focused task based on disassembly. Other skills such as Tasting or Food Costing can be taught within this framework.

Lesson plans

Lesson plans have been created for each section and these plans include details of **knowledge and skills**, homework ideas and what will be **assessed** and how.
The lesson plans suggested are for one hour lessons over a period of 4-8 sessions. This is based on current practice for most food technology teachers.

Simple beginnings
The activities start simply with "designing a muesli" and **progress** in **developing competence** of food handling skills. The ideas are deliberately prescriptive in the task of disassembly, but the design and make activities can be very open ended. So, the hot bread snack can end up with a design of any shape, size and topping.

Differentiation
Differentiation by outcome
The tasks in the book start simply and become progressively more complex. Students can produce better quality designs as they improve their knowledge and skills. As they become more skilled at writing **detailed specifications** and **quality checklists** so their work will improve, and they will achieve differentiation by outcome. Teachers can help students progress by encouraging the development of these skills.
Diffentiated questions.
The questions become more complicated as the tasks progress. Where Further Work is suggested, this involves more detailed research and investigation.
Progression in knowledge and skills
The charts on progression show how, through a series of tasks, students can progress in their knowledge and understanding of being able to **disassemble a food product, designing for quality, and writing a specification.** The charts can be adapted for assessment purposes.
The work has been trialled with teachers and students and their criticisms and ideas incorporated into the text.

Copyright Ridgwell Press, PO Box 3425, London SW19 4AX. This page may be photocopied by purchasing institution only.

National Curriculum Orders

Key Stage 3 Programme of Study

Pupils should be taught to develop their design & technology capability through combining their **Designing and Making skills** with **Knowledge and understanding** in order to design and make products.

1. Pupils should be given opportunities to develop their design & technology capability through:

a assignments in which they design and make products focusing on different contexts and materials, and including the use of:

1 resistant materials

2 compliant materials and**/or food**

Taken together, these assignments should include work with control systems, eg electrical, electronic, mechanical, pneumatic, and structures.

b Focused practical tasks in which they develop and practise particular skills and knowledge.

c Activities in which they investigate, disassemble and evaluate familiar products and applications.

2. Pupils should be given opportunities to:

a Work independently and in teams.

b Apply skills, knowledge and understanding from the programmes of study of other subjects, where appropriate, including art, mathematics and science.

Pupils should be taught to:

3. Designing skills

a **Identify appropriate sources of information that will help with their designing.**

b **Use design briefs to guide design thinking.**

c **Develop a specification for their product.**

d **Consider the needs and values of intended users, and develop criteria for their design to guide thinking and form a basis for evaluation.**

e **Generate design proposals that match stated design criteria, and modify proposals to improve them.**

f Consider the aesthetics, function, safety, reliability and cost of their designs.

g Take account of the working characteristics and properties of materials and components when deciding how and when to use them.

h **Prioritise and reconcile decisions on materials and components, production, time and costs within design proposals.**

i Take account of the restrictions imposed by capacities and limitations of tools and equipment.

j Explore, develop and communicate their design ideas by modelling their ideas in an increasing variety of ways, including the use of IT.

k Develop a clear idea of what has to be done and propose an outline plan, which includes alternative methods of proceeding if things go wrong.

l **Evaluate their design ideas as these develop, bearing in mind the users and the purposes for which the product is intended, and indicate ways of improving their ideas.**

4. Making skills

a Use a range of processes to shape and form materials, including forming by wastage, attachment, adhesion and combining.

b Select materials, tools and equipment appropriate to the task.

c Select and use appropriate methods of shaping and forming materials accurately.

d Join and combine additional materials and components accurately in temporary and permanent ways.

e Use construction kits that offer a wide choice of interconnections and components.

f Interconnect a variety of components to achieve functional results.

g Apply a range of finishing techniques appropriate to the materials being used, taking into account the purposes to which the finished products will be put.

h Make products in quantity, using techniques to ensure consistency.

i Develop a clear idea of what has to be done and propose an outline plan that includes the materials, equipment and processes to be used, and suggests alternative methods of making if first attempts fail.

Statements arrowed and in bold are areas covered by ideas in this book

j **Evaluate their products as these develop, including testing performance against specified criteria.**

k **Implement improvements they have identified and take ongoing action to ensure that their products meet the specification and their original intention.**

Knowledge and understanding

Pupils should be taught:

5 Materials and components

a **To consider the physical and chemical properties of materials and to relate these properties to the ways materials are worked and used.**

b That materials can be classified according to their properties and behaviour, and the major classifications within the material categories they are using, eg thermoplastics and thermosets.

c That materials can be combined, processed and finished, in order to create more useful properties and desired aesthetic effects, eg combining different ingredients to create products with different sensory characteristics.

d That heat treatment and the combining of materials alter working and performance characteristics.

e That some materials can be formed by pressing or casting which are important techniques for producing.

6. Systems and control

a To design, use and interconnect simple mechanical,electrical, electronic and pneumatic systems and sub-systems.

b How to interconnect mechanisms to achieve different kinds of movement.

c To use electrical switches to control devices.

d To use sensors in switching circuits.

e That systems have inputs, processes and outputs, and to recognise these in existing products and products they have made.

f The importance of feedback, and how it can be used to ensure the correct functioning of mechanical, electrical or electronic systems.

g To analyse the performance of systems, in order to check that they are working effectively.

7. Structures

a To recognise and use structures in their products.

b That excessive loads can cause structures to fail by bending, buckling and twisting.

c To use simple tests to determine the effects of excessive loads.

d To devise suitable methods to reinforce their structures and relate these and other techniques to familiar structures.

e To understand that forces, such as compression and tension, produce different effects, and to take account of these in making their products.

8. Products and applications. Pupils should be taught to investigate, disassemble and evaluate a wide range of products and applications, in order to learn how they function, and relate products to:

a Their intended purpose.

b The choice of materials and components, and the ways in which they have been used.

c The processes used to produce them.

d The scientific principles applied.

e The views of users and manufacturers.

f A range of alternative products.

9. Quality - Pupils should be taught to distinguish between quality of design and quality of manufacture, in order to identify, and use criteria that help them judge the quality of a product, including:

a How far it meets a clear need.

b Its fitness for purpose.

c Whether it is an appropriate use of resources.

d Its impact beyond the purpose for which it was designed, eg on the environment.

10. Health and safety - Pupils should be taught further knowledge and understanding of health and safety as designers, makers and consumers, including:

a Taking responsibility for recognising hazards in a range of products, activities and environments with which they are familiar.

b Using appropriate information sources to assess the risks, both immediate and cumulative.

c Applying their knowledge and take action to control the risk to themselves and to others.

Copyright Ridgwell Press, PO Box 3425, London SW19 4AX. This page may be photocopied by purchasing institution only.

Progression in being able to disassemble a food product

This chart shows how students can build progression into learning how to disassemble a food product.

Be able to describe the looks and taste of the food, with labelled drawings.

⇩

Weigh and measure the parts of the food - for example, weigh the pastry and filling.

⇩

Be able to work out how it is made and the order of making (making process) - for example, pie has pastry base, then filling put in and then top pressed on.

⇩

Know which ingredients are used in the product by using the list on the packaging label.

⇩

Know what the ingredients do in the food (their functions) - for example, flour is used for the pastry and apple for the filling.

⇩

Show the limits in being able to copy a ready-made food - for example, some ingredients are not available to buy.

⇩

Find out the cost, portion size, shelf life, storage and cooking instructions. Be able to apply this knowledge when designing other food products and know the limitations.

⇩

Be able to recognise the quality standards used in the product's manufacture.

⇩

Be able to write a specification for the product showing target group, ingredients, cost, nutrition information, shelf life, size, shape, taste, portion size and quality.

Copyright Ridgwell Press, PO Box 3425, London SW19 4AX. This page may be photocopied by purchasing institution only.

Progression in designing for Quality

This chart shows how students can build progression into learning about quality control of food product design

Student should:

Be able to describe in simple terms the quality of the food product - how it should look and taste.

⇩

Add details to describe how to meet these quality standards - for example, crispy garlic bread needs to be properly cooked.

⇩

Show how quality controls can be linked into the making process.

⇩

Be able to evaluate the finished product and see if quality controls have been met - if not, be able to improve next time.

⇩

Understand why quality control is important to the food industry if safe and successful food products are to be made.

⇩

Know that if quality controls are not in place, you end up with a poor quality end product.

⇩

Know about HACCP and food temperature control in the food industry.

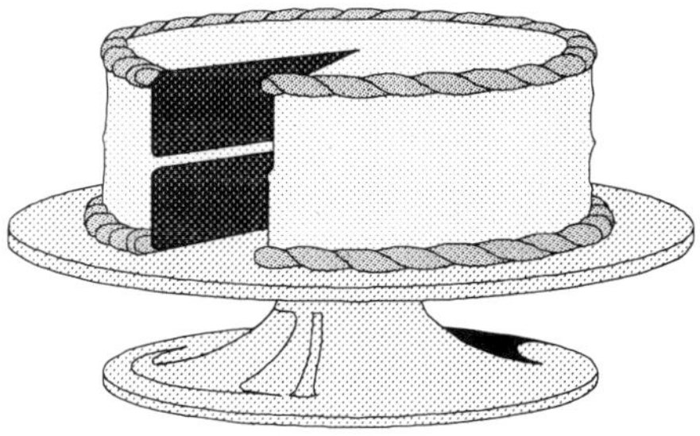

Copyright Ridgwell Press, PO Box 3425, London SW19 4AX. This page may be photocopied by purchasing institution only.

Progression in writing a specification for a food product

Writing a specification:

Specification for food products should cover:
* size, shape and weight
* shelf life
* nutritional composition
* appearance and sensory characteristics
* cost
* hygiene standards.

Food must be safe and clean to eat, so show safe and hygienic work practice.

Progression in writing a specification for a food product:

Know the target group and their needs. Show how the product specification meets these needs, eg. vegetarians - no meat, fish, chicken.

⇩

Draw a star profile to show the specification. Describe the looks, flavour and texture.

⇩

Make a simple list of ingredients and weights, with method for making.

⇩

If appropriate include:
a detailed list of ingredients (recipe formulation)
serving or portion size
size and shape using drawings
preparation details (making process).
costing,
how long the product will last,
details of temperature control in storage,
nutrition information - back up claims for "healthy food"
hygiene and safety
cooking instructions
label information
packaging information
quality of ingredients required.

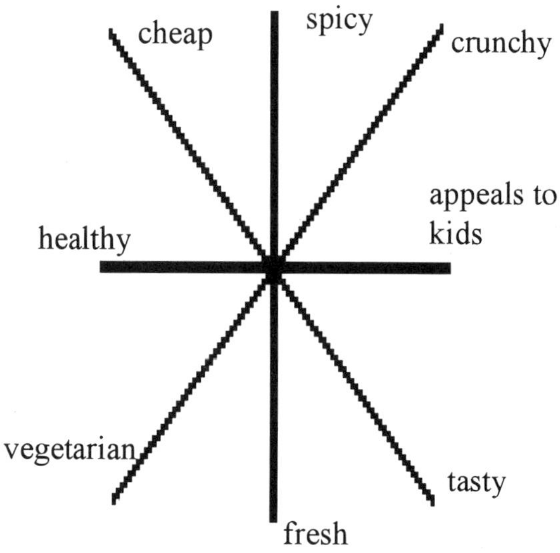

Star profile of vegetarian meal for children

This chart shows the progression in writing a simple specification for the recipe and method for garlic bread:
Garlic bread recipe - breadstick, butter, garlic and salt
Method
Mix butter, garlic and salt. Slice bread and spread on butter. Bake until cooked.
This is a more complex specification for the recipe and method:
Garlic bread recipe
200g baguette, 50g butter, 3 cloves crushed garlic, large pinch of salt
Method
Cut bread into 2.5 cm slices half way down the bread.
Mix butter, garlic and salt until smooth.
Spread butter evenly between slices and on top of bread. Bake 200° C, Gas Mark 6 for 15 minutes until crisp.

 Copyright Ridgwell Press, PO Box 3425, London SW19 4AX. This page may be photocopied by purchasing institution only.

Looking at muesli lesson plan

Time allowance
Series of 4, 1 hour lessons.

Lesson Plans

Disassemble a muesli to find what it is made from.

⇩

Plan design for own muesli.

⇩

Make the muesli and test it.

⇩

Discuss results and quality control.

What are you teaching pupils? - teaching intention?
To learn how to take a very simple food apart.
How to write a specification.
Design and make an easy-to-assemble food to test against the specification.

How to teach disassembly
This is a simple way to start design work on food. Look closely at a packet of muesli and work out what ingredients have been used (disassembly) and what their purpose (function) is in the recipe. Use this information to write a specification for a muesli. Then make and taste the muesli and see if it meets the specification.
Build in progression to this activity by choosing ingredients according to their function - for example, sweetness - sultanas. Also think about the quality of the ingredients to be used.

How will it be assessed?
Find out how much information groups of students have got from looking closely at the muesli.
How detailed is the specification for their own design?
Can they test to see if their muesli meets the specification?

4 lesson plans and assessment

Lesson	Lesson Plan	Knowledge and skills	Homework	What will be assessed and how
1	Investigate and disassemble a muesli.	How to get information by taking food apart (disassembly). How to write specifications.	Write the specification for the muesli.	Homework with mark or grade.
2	Plan to make your own muesli.	Planning your own design based on disassembly.	Write out plan of work and ingredients.	Homework with mark or grade.
3	Make the muesli and taste it.	Following plans and working to own design.	How well did it go? Write up practical results.	Evaluating practical. Homework.
4	Discuss results and quality control.	Quality of food products.	Evaluate work. What new knowledge and skills did you learn?	Discussion - how well did it go and what needs improving?

Copyright Ridgwell Press, PO Box 3425, London SW19 4AX. This page may be photocopied by purchasing institution only.

Hot bread snack lesson plan

Focused task - disassembly of a food product
The charts show how the disassembly of a hot bread snack such as garlic bread can be built into a series of 4 lessons.

Lesson Plans

Disassemble a range of hot bread snacks to find out how they are made.

⇩

Plan design for own hot bread snack.

⇩

Make and taste their own design.

⇩

Compare costs, taste and nutrition. Complete work.

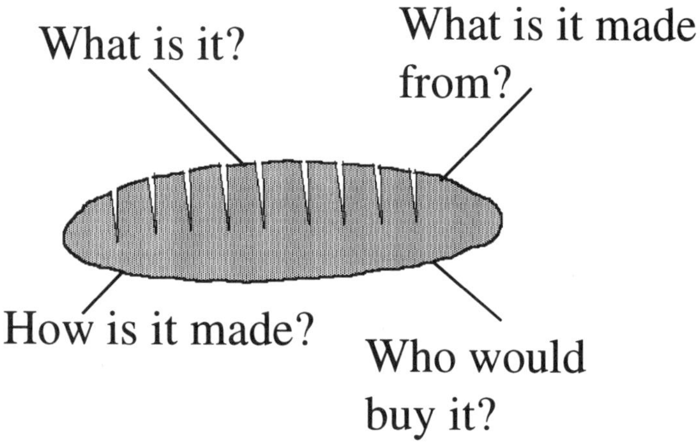

This chart shows the knowledge and skills associated with the lessons and simple assessment procedures.

Lesson	Lesson Plan	Knowledge and skills	Homework	What will be assessed and how
1	Investigate and disassemble different garlic breads and hot bread snacks.	How to get information by taking food apart (disassembly). How to write specifications.	Write the specification for the garlic bread or bread snack.	Homework with mark or grade.
2	Choose ingredients for own design of hot bread snack.	Planning their own design based on disassembly.	Write out plan of work and ingredients.	Homework with mark or grade.
3	Make the bread snack and taste it.	Following plans and working to own design.	Work out the cost of the bread snack and write up practical results.	Observing practical outcome. Homework.
4	Complete evaluation of bought and made products	Comparing costs, looking at quality of food products.	Evaluate work. What new knowledge and skills did you learn?	Discussion - how well did it go and what needs improving?

Copyright Ridgwell Press, PO Box 3425, London SW19 4AX. This page may be photocopied by purchasing institution only.

Fruit pie lesson plan

Focused task - Disassemble a food product, learn about quality control, specifications and practise skills with pastry and pie making.

Time allowance
Series of 5-9, 1 hour lessons.
The suggested lesson plan can take 5 weeks or last longer if a Design and make task is included.

Lesson plans

Disassemble the fruit pies to find how they are made and what from.

⇩

Learn how to make pastry, write specification. Design own pie.

⇩

Make own fruit pie. Build in quality controls.

⇩

Taste and compare home-made with bought pie.

⇩

Compare costs, labels and nutrition.

⇩

Design and make for 4 weeks.

What are you teaching the pupils?
To learn how to take a more complex food apart (disassemble).
To compare costs and quality of home-made and ready-to-eat food products.
To carry out a design-and-make task using these skills.

How to teach it
Look at a range of fruit pies. Work out how they are made. Design and make your own version. Compare costs and quality.

How will the work be assessed?
Find out how much information the student has got from examining the product closely.
How detailed is the home-made pie specification?
When the pie is made, how closely does it meet the specification?
Have the quality controls been met?
How successful is the final pie in terms of quality?

Design and make task
Task - to find out which is better quality - home-made or ready-to-eat food products. Design and make a food product to test out this task.

Detailed lesson plans and homework information are given on the following page.

Copyright Ridgwell Press, PO Box 3425, London SW19 4AX. This page may be photocopied by purchasing institution only.

Course title - Looking at fruit pies to get design information

Lesson	Lesson Plan	Aim of lesson for focus	Homework	What will be assessed and how
1	*Take apart and taste ready-to-eat fruit pies. Look closely at packaging, ingredients, size, shape, decoration and cost.*	*Understand how to get information by taking food apart - disassembly. Learn about specifications.*	*Write the specification for the fruit pies.*	*Homework with mark or grade.*
2	*Demonstration of pastry making. Write a specification for home-made pie.*	*Understand skills needed for pastry and pie making. Reinforce work on specifications.*	*Write specification for home-made pie and write time plan for practical.*	*Homework with mark or grade.*
3	*Make the home-made pies to the specification. Freeze some pie for tasting.*	*Following a specification and time plan.*	*Evaluate practical work. Work out the cost of home-made pie.*	*Self assessment of practical work.*
4	*Carry out a tasting to compare the ready-to-eat pie with the home-made.*	*Using tasting methods which will obtain useful information to compare the pies.*	*Write up the results of the tasting with an evaluation.*	*Homework with mark or grade.*
5	*Compare costs, nutritional information, look at labels.*	*Cost and nutritional comparison, knowledge of labelling information.*	*Evaluate results of work to date.*	*Discuss what student knows about specification and what improvements are needed.*
6	*Design and make assignment for 4 weeks to compare the value of ready-to-eat food with home-made.*	*Using knowledge and skills to develop own design ideas.*	*Students plan homework needs.*	*National Curriculum Assessment.*
7	"	"	"	"
8	"	"	"	"
9	*Complete work*		*Evaluation of task.*	*Discussion:- How well did it go? Where do we go next?*

Children's food lesson plan

Task - to design a food product for an under 10 year old.

Time allowance
Series of 6, 1 hour lessons.

Lesson plan - the chart below shows the 6 detailed lesson plans.

What are you teaching?
To learn how to look at a range of food products and see how they meet the needs of the target group.
More detailed investigations to include packaging, labelling and nutritional information.

How to teach disassembly
Collect a range of products suitable for the target group - children's food for children.
Look carefully at the packaging design, labelling and then taste the contents.
Come up with a specification based upon results.

How will it be assessed?
Students can carry out a lot of self assessment for this task since they will have developed competency in this area.
Specifications should be aimed at the needs of the target group - children. Students can arrange for children to taste their food designs.

Lesson	Lesson Plan	Knowledge and skills	Homework	What will be assessed and how
1	Investigate and disassemble a range of foods aimed at the target group (children). Write a specification	How to get information by taking a wide range of food apart (disassembly). Writing specifications.	Write the specification.	Homework with mark or grade.
2	Research needs of target group - use questionnaire. Plan 2-3 ideas to cook / trial to meet brief/ specification.	Creating questionnaire. Designing to specification.	Carry out questionnaire and present results.	Homework with mark or grade.
3	Make 1-2 food products. Taste and test.	Making food to plan.	Evaluate work.	Self evaluation.
4	Make 1-2 food products. Taste and test.	"	"	"
5	Choose and develop the most suitable product. Decide upon packaging and labelling.	Modify ideas. Packaging and labelling.	Design packaging.	Homework with mark or grade.
6	Complete work.		Evaluate task	Discuss ideas - did it go to plan, and what needs improving.

Copyright Ridgwell Press, PO Box 3425, London SW19 4AX. This page may be photocopied by purchasing institution only.

Lesson	Lesson plan	Aim of lesson for focus	What will be assessed and how

Copyright Ridgwell Press, PO Box 3425, London SW19 4AX. This page may be photocopied by purchasing institution only.

Useful forms to photocopy

Food Product Specification

Size

Shape

Weight

Looks

Taste

Tick a box ☑ for each statement

Quality Report

	Yes	Needs improving	No
Worked safely	☐	☐	☐
Worked hygienically	☐	☐	☐
Stored and cooked food at safe temperatures	☐	☐	☐
Used knowledge and skill	☐	☐	☐
Worked to plan	☐	☐	☐
Worked efficiently	☐	☐	☐
Cleared up	☐	☐	☐
Presented work on time	☐	☐	☐
Made a quality, tasty food product	☐	☐	☐

Copyright Ridgwell Press, PO Box 3425, London SW19 4AX. This page may be photocopied by purchasing institution only.

Food Product Analysis

Name of food product

Drawing

What does it look like?	What does it weigh?

Copyright Ridgwell Press, PO Box 3425, London SW19 4AX. This page may be photocopied by purchasing institution only.

Food Product Analysis

Taking apart (disassembling) a food product to get design information

How does it look?

How does it taste?

How is it made?

Other useful information

What ideas does it give you for food product design?

Copyright Ridgwell Press, PO Box 3425, London SW19 4AX. This page may be photocopied by purchasing institution only.

Useful forms to enlarge and photocopy

Food product analysis			
Ingredients	**Weight**	**Appearance of ingredient**	**Functions of ingredient**
Total weight = ***Cost =***			

Food Product Analysis

What is it?	**What is it made from?**
Who would buy it to eat?	**How is it made?**

Copyright Ridgwell Press, PO Box 3425, London SW19 4AX. This page may be photocopied by purchasing institution only.

Making for Quality

Useful forms to enlarge and photocopy

	Making process	Quality Checklist

Time	**Order of Work**	**Quality Control**

Copyright Ridgwell Press, PO Box 3425, London SW19 4AX. This page may be photocopied by purchasing institution only.

Food product specification

My target group is:..

Their special needs are: ....

...

...

This is what I want the food to be like:

Steps to show how the food is made:

The ingredients needed are:

Ingredients	Quantity

Evaluation when the food is made - does it meet the specification?

Copyright Ridgwell Press, PO Box 3425, London SW19 4AX. This page may be photocopied by purchasing institution only.

Glossary of terms

Assignments - tasks to design and make products focusing on different contexts and materials.
"Best before" date - shows the consumer the period when the food product is in its best condition if kept according to the manufacturer's instructions.
British Standards - standards which set requirements for products and product performance and reliability. BS 5750 (1979,1987) relates to Quality Management Systems.
Cold spots - areas in food not heated properly. Happens in microwave ovens when an area cannot be reached by microwaves.
Critical control points - the step in the preparation of food which has to be carried out correctly to make sure that a hazard is removed or reduced to a safe level.
Critical temperature - the temperature needed to get the required effect - for example - food is cooked to a high enough temperature to kill bacteria or cause eggs to set.
Cross contamination - the transfer of a substance (e.g. microbes) from one area to another - one food can be contaminated with bacteria from another.
Design brief - a concise statement which sets out the problem to be solved and gives the context within which possibilities are discussed.
Design criteria - the desirable performance characteristics of the product to be designed.
Design specification - a detailed statement derived from the design brief which describes the characteristics of the product to be designed.
Disassemble - take apart a product to find out how it is made and what it is made from.
Environmental Health Officer (EHO) - the enforcement officer at local government level who covers public health such as the hygiene of food premises and food safety.
Focused practical task - a task which uses and develops particular skills and knowledge such as disassembly of food.
Food processing - the way food is changed to turn it into something else. So food is preserved by freezing, and made safe by cooking.
Food safety checks - looks at temperature control, microbial checks and cross contamination.
Hazard - anything which can cause harm to a consumer. A hazard may be biological such as salmonella in chicken, physical such as glass in food or chemical such as chemicals in food.
Hazard Analysis and Critical Control Point (HACCP) - a system for caterers and food manufacturers used to control food safety problems.
Hygiene - making the work place and the workforce work in a clean way which is safe for health.
MAFF - Ministry of Agriculture, Fisheries and Food.
Quality assurance - a system which aims to ensure that quality is inbuilt by means of written procedures and communications.
Quality assurance procedures - checks carried out at different stages of production to ensure that the required standards are maintained. Tests can be undertaken by sampling.
Quality control - "operational techniques and activities that are used to fulfil requirements for quality" (BS 4778).
Quality food checks - look at appearance of raw and cooked food, flavour, texture, packaging and labelling.
Quality management - "a systematic way of guaranteeing that organised activities happen in the way they are planned." (P Crosby Quality is free, 1979, McGraw Hill)
Quality of design - creating a design which is easy to make with a clear specification and production process. Quality design reduces costs and increases profits, as well as meeting the needs of consumers.
Quality of manufacture - building in quality controls and checks into the way the product is made.
Quantitative Declaration of Characterising Ingredients (QUID) - This is a proposed system for use on food labels. Certain ingredients which characterise a food stuff will be declared to show how much is present. In a chicken and ham pie, consumers will know how much chicken and how much ham is present. The system is under discussion (Dec 1994) and is proving complicated to implement. Too much detail of the content of a food product will give away the details of the recipe. QUID does provide useful information for consumers to judge which pie contains the most chicken, or which tomato soup the most tomatoes.
Salmonella - a family of bacteria which causes diseases including food poisoning.
Shelf life - the shelf life of a product is the period of time over which any particular food product can be expected to last under the recommended conditions of storage.
Specifications - for food products these should cover details of size, shape, weight, shelf life, nutritional composition, appearance and sensory characteristics, cost and hygiene standards.
Trading Standards Officer - the enforcement officer at local government level who covers commercial law such as weights and measures and the correct labelling of food products.

Answers to questions on Fruit Pie labels P20-21

Nutrition
1. Sainsbury's label
2. Lyons. This pie may be larger in size than the others, but also more fat, which is energy dense, may be used in the recipe.
3. Sainsbury's pies show the amount of sugar - 16g per pie = 3 teaspoons.

Ingredients
1. for pastry - a) wheatflour, fat, sugar (plus additives) for filling - b) fruits and sugars and gels
2. To provide more information for consumers on the quality of the product.
3. sweetening pies - sugar, glucose syrup, dextrose, fats - b) hydrogenated vegetable fat or oil.
4. additives - a) preservatives - potassium sorbate, b) stabilisers - sodium alginate, c) gelling agents - pectin, sodium alginate, d) colours - lutein, anthocyanins, e) acidity regulators - sodium citrate.
5. The ingredients are not readily available, processes not available and additives can't be used.

Copyright Ridgwell Press, PO Box 3425, London SW19 4AX. This page may be photocopied by purchasing institution only.

Index

The Chase High School, Geraldine Road, MALVERN, Worcs. WR14 3NZ. Tel: 01684 57[illegible]96 Fax: 01684 566643

Useful resources

Books

Tasting and Testing, Ridgwell, Ridgwell Press, ISBN 09521645 07
Nutrition and Food Design, Ridgwell, Ridgwell Press, ISBN 09521645 15
Food Coursework, Ridgwell, Ridgwell Press, ISBN 09521645 31
Food Temperature Control Workcards, Ridgwell and Kitchen, Ridgwell Press.
All above available from Ridgwell Press, PO Box 3425, London SW19 4AX.
Food for Thought , Key Stage 3, Cambridge University Press, ISBN 0521406366
ATaste of Quality - An introduction to the importance of quality in modern manufacture. Free resource. McDonald's Restaurants Ltd., 11-59 High Road, East Finchley, London N2 8AW.

Videos with information on product development

Some Catch - A video for 14-16 year olds designed to raise commercial areas such as product development, manufacturing and marketing as well as fish as a food source. Cost £10.00 including p&p. Available from Sea Fish Authority, Education Information Service, 18 Logie Mill, Logie Green Road, Edinburgh EH7 4HG.

Look at the label (New 1993 version), MAFF, running time 16 minutes
This is a very jolly video suitable for key stages 3 and 4, which describes how food labels show nutritional value, storage time, cooking details and much more. The video producers have thought carefully about their target audience and made a rather boring subject quite fun and memorable.
Available on free loan from CFL Vision, PO Box 35, Wetherby, West Yorkshire, LS23 7EX, Tel 0937 541010.

What shall we eat tonight? 14 minutes, free from MLC, PO Box 44, Winterhill House, Milton Keynes, MK6 1AX
A video which shows the five stages of product development: Consumer research, the product, developing and testing, marketing strategy and product launch.

Biscuit Making
A really good video showing biscuit making in an Australian factory. The film shows designing and making on the large scale, which is excellent for the new Design and Technology orders. Since the factory makes many types of biscuit, you see how biscuit dough is shaped with cutters and moulds. There is useful information on quality control - every 10 minutes the product is checked for moisture, colour and weight. Cost £27 plus VAT and postage. Classroom Video, Darby House, Bletchingley Road, Merstham, Redhill, Surrey, RH1 3DN, Tel 0737 64 2880.

Extracts from National Curriculum Orders, Design and Technology Key Stage 3 reproduced by permission of the Controller of Her Majesty's Stationery Office.

Food Activities - Disassembly and Quality Control - ISBN 09521645 5 8

Copyright Ridgwell Press, PO Box 3425, London SW19 4AX. This page may be photocopied by purchasing institution only.